WILD STEM

STEM in the FOREST

Megan Borgert-Spaniol

Checkerboard Library

An Imprint of Abdo Publishing
abdobooks.com

abdobooks.com

Published by Abdo Publishing, a division of ABDO, PO Box 398166, Minneapolis, Minnesota 55439.

Printed in the United States of America, North Mankato, Minnesota
102023
012024

THIS BOOK CONTAINS
RECYCLED MATERIALS

Design: Denise Hamernik, Mighty Media, Inc.
Production: Mighty Media, Inc.
Editor: Anna Anderhagen
Cover Photographs: Maskot/Getty Images
Interior Photographs: Dmytro Gilitukha/Adobe Stock, p. 17; Gonzalo Keogan/Shutterstock Images, p. 21; Gorodenkoff/Shutterstock Images, p. 23; HighwayStarz/Adobe Stock, p. 9; Iam_Anuphone/Shutterstock Images, p. 5; Ivan Kurmyshov/Shutterstock Images, p. 27; Khlongwangchao/iStockphoto, p. 13; Mark Johnston/AP Images, p. 11; Matthew Brown/AP Images, p. 7; Mighty Media, Inc., pp. 14–15, 24, 25; Orhan Cam/Shutterstock Images, p. 29; Szymon Bartosz/Adobe Stock, p. 19
Design Elements: Lorthois Yuliya/Shutterstock Images; supanut/Adobe Stock

Library of Congress Control Number: 2023939338

Publisher's Cataloging-in-Publication Data
Names: Borgert-Spaniol, Megan, author.
Title: STEM in the forest / by Megan Borgert-Spaniol
Description: Minneapolis, Minnesota : Abdo Publishing, 2024 | Series: Wild STEM | Includes online resources and index.
Identifiers: ISBN 9781098291952 (lib. bdg.) | ISBN 9781098278854 (ebook)
Subjects: LCSH: Science--Study and teaching--Juvenile literature. | Technology--Study and teaching--Juvenile literature. | Engineering--Study and teaching--Juvenile literature. | Mathematics--Study and teaching--Juvenile literature. | Forests and forestry--Juvenile literature. | Trees--Juvenile literature.
Classification: DDC 577.0--dc23

CONTENTS

WILD FOREST

Do you love the smell of cedar and moss? Do you dream of dirt paths that wind among towering trees? If so, you might enjoy a career that takes you to the forest!

Many people work in forests. Forest rangers care for public forests and their visitors. Urban foresters improve cities with tree-filled spaces. Forestry workers grow and manage forests, while loggers harvest forests for their lumber. Wildland firefighters combat and prevent forest fires.

What kind of job would you like to have? There's only one way to find out. Put on your hiking boots and grab a compass. It's time to explore science, technology, engineering, and math (STEM) in the forest!

STEM
Science
Technology
Engineering
Math

Forests remove air pollution, filter water, and provide homes for wildlife. More than 1.6 billion people depend on forest resources.

FORESTRY WORKERS

Forests are valuable landscapes. Forests are home to more than 80 percent of land-dwelling plants and animals. It takes a lot of work to maintain forests, but forestry workers are ready for the task.

Forestry is the science of growing, managing, and conserving forests. Many forestry workers **specialize** in one of these areas. For example, some forestry workers study soil health for planting trees. Others oversee logging operations when trees are ready to harvest. Some protect forest **habitats**.

An important part of forestry is wildfire prevention. Forestry workers remove **vegetation** that can fuel wildfires. They clear brush, **prune** branches, and conduct controlled burns.

A forestry worker loads highly flammable eucalyptus logs onto a truck near Holy City, California. He is part of a wildfire prevention crew removing dry vegetation.

A variety of tools help forestry workers do their jobs. Clinometers measure the height of trees. **Diameter** tape measures the thickness of trees. And bark gauges measure the thickness of trees' bark. All these measurements help workers **determine** how much lumber a tree will produce.

More advanced technologies gather data on entire forests. One example is a remote **sensor** called lidar. Lidar sends laser light from an aircraft toward the ground. The light reflects off the trees and returns 3D data on tree height and **density**. This data helps forestry workers track tree growth, **monitor** forest health, and more!

Forestry workers decide which trees to plant and how many. They also track tree growth and inspect new trees for disease.

FOREST RANGERS

Have you ever been to a national park or forest? You will find park or forest rangers at these public lands. Forest rangers take care of the land and the wildlife it supports. They also help visitors who come to experience it.

One of a forest ranger's primary duties is to protect the forest. Forest rangers patrol the forest and enforce park rules. For example, they might stop visitors from littering or damaging trees. They also ensure visitors follow laws for hunting, fishing, and camping.

Forest rangers also protect wildlife in the forest. They control pests, weeds, and **invasive species**. They direct visitors away from animal **habitats**. Sometimes, rangers care for animals that are sick or injured.

A forest ranger patrols the Timpooneke Trail on Mount Timpanogos in Utah.

In addition to protecting the forest and wildlife, forest rangers serve visitors. They maintain trails to keep them safe for hikers. They lead guided tours to teach visitors about the forest. They also create educational programs about wildlife, safety, and wildfire prevention.

Forest rangers use many tools to do their jobs. They rely on land vehicles, snowmobiles, and sometimes helicopters to get around quickly. They **monitor** wildlife with the help of GPS tracking devices. In remote areas, they use radio communication to send messages to other rangers. Rangers also use technology to create live streams, virtual hikes, and smartphone walking tours for visitors.

If a visitor gets lost or injured, forest rangers may lead paramedics on a search and rescue mission to bring the visitor to safety.

GET WILD!

ANIMAL TRACKS

Forest rangers monitor wildlife by identifying animal tracks. Make a cast of an animal track when you hike in the forest!

WHAT YOU NEED:

- ✔ measuring cup
- ✔ sealable plastic bag
- ✔ plaster of paris
- ✔ water bottle
- ✔ water
- ✔ backpack
- ✔ stick
- ✔ old toothbrush
- ✔ permanent marker

WHAT YOU DO:

1. Pour 2 cups (473 mL) plaster of paris into the sealable bag. Pour at least 1 cup (237 mL) water into a water bottle.

2. Pack the sealable bag, water bottle, and measuring cup in a backpack that you'll wear on your forest hike. Look for animal tracks!

3. When you find an animal track, pour 1 cup (237 mL) water into the bag of plaster of paris. Seal the bag and knead it until the plaster is smooth.

4. Use a stick to carve a dam around the track. Pour enough plaster to completely cover the track. Let it sit for 30 minutes.

5. Remove the cast from the ground. Take it home and let it harden overnight.

6. Use a toothbrush to clean any dirt off the cast. Use a permanent marker to record the date, type of animal track, and where it was found on the back of the cast.

LOGGERS

The US government protects trees and land in public national forests. But many privately owned forests are maintained to produce lumber. Lumber is used to construct buildings and furniture. It is also used to make paper, musical instruments, and many other products.

Tree harvesting is the process of cutting down trees, trimming them into logs, and transporting them out of the forest. This process is also known as logging. The people who do this work are called loggers.

Loggers **specialize** in certain parts of the tree harvest. For example, some loggers focus on the work of cutting down trees. They are known as fellers or fallers. Other loggers, called buckers, cut the trees into logs.

Besides wearing a logger's helmet, a logger wears earmuffs and a face shield to protect their ears and eyes.

Loggers must remove logs from the logging site and load them onto trucks. Many loggers train to operate heavy machinery, such as a crane or tractor. Logs also need to be inspected and measured to **determine** their value. The loggers who do this task are called log graders and scalers.

The lumber produced by logging is a valuable resource. But protecting forest ecosystems is also important. **Sustainable** logging practices consider a forest's health and how much lumber it can produce. For example, loggers might harvest trees from a few small forest areas instead of an entire forest. Sustainable logging ensures our forests are healthy for many years to come!

A log grader measures the diameter of a tree trunk using a tree caliper. Her calculations will determine the value of the log and how much lumber it will produce.

WILDLAND FIREFIGHTERS

Forests are full of brush, dead branches, and other flammable material. This material makes forests **vulnerable** to wildfires, especially in dry weather conditions. Wildfires result from various activities and events, including unattended campfires and lightning strikes. At the first sign of a wildfire, wildland firefighters spring into action!

When wildland firefighters arrive at the fire, they build a fire line. This fire line is a strip of land cleared of brush and other fire fuel. Firefighters create fire lines using chain saws, axes, and rakes. Smokejumpers are **specialized** wildland firefighters trained to parachute into remote areas. Smokejumpers contain small wildfires before they get out of control.

After a fire line is created, a wildland firefighter extinguishes the fire with water.

Another group, known as hotshots, handles the most challenging wildfire conditions. Hotshot crews battle steep **terrain**, high winds, and fast-moving flames. Among their safety gear are fire shelters. These tent-like structures withstand temperatures up to 2,000 degrees Fahrenheit (1,093°C) for a short time. Firefighters **deploy** them if they have no other way to escape a blaze.

When they are not fighting wildfires, wildland firefighters also work to prevent them. They use hand tools and machines to thin out overgrown areas. Thinning out **vegetation** removes fire fuel. It also improves forest health by creating space for new growth!

Wildland firefighters must be in excellent physical shape. They often carry more than 50 pounds (23 kg) of gear over long distances.

GET WILD!

FIGHTING FIRE

A chemical reaction between oxygen, heat, and fuel creates fire.

In this project, see what happens to fire when you take away its oxygen.

WHAT YOU NEED:

- ✔ small candle
- ✔ matches
- ✔ funnel
- ✔ measuring cups and spoons
- ✔ baking soda
- ✔ empty bottle with a cap
- ✔ vinegar
- ✔ glass jars of different sizes

WHAT YOU DO:

1. Have an adult help you light the candle. Use the funnel to pour 2 tablespoons (30 mL) baking soda into the empty bottle. Add ¼ cup (59 mL) vinegar to the bottle. Quickly cap the bottle.

2. Wait for the solution in the bottle to stop bubbling. Hold the bottle sideways with the top near the candle. Remove the cap but don't let the liquid spill out.

3. Observe what happens. The baking soda and vinegar mixed together produce carbon dioxide. When this gas flows out of the bottle, it pushes away the oxygen that the candle needs to burn.

4. Have an adult help you relight the candle. Place a jar upside down over the candle. Observe what happens. The jar seals off the candle from oxygen. Once the candle uses all the oxygen in the jar, it burns out.

5. Repeat step 4 with a few jars of different sizes. Does the size of the jar affect how long the candle burns?

URBAN FORESTERS

Research shows that trees help improve humans' physical and mental health. But most humans don't live in forests. More than half of the world's population lives in cities! That's why cities must have urban forests.

Urban forests take the form of parks, public gardens, and nature preserves. Trees pull carbon dioxide from the air, which helps clean it. Trees release oxygen for us to breathe. Trees also reduce urban heat by providing shade.

Urban foresters plan, grow, and maintain urban forests. They use computer **software** to map urban forests. Urban foresters work with engineers, landscape architects, and urban planners. Together, they work to give all urban communities access to the benefits of trees.

Urban forestry and architecture meet in these residential towers in Milan, Italy. The towers are called *Bosco Verticale*, which means "vertical forest" in Italian.

A CAREER IN THE FOREST

Do you have a passion for STEM in the forest? There are many career paths to choose from. Each path leads to a wide range of opportunities.

A forestry worker might work for the government or a university. A forest ranger could work at a state or national park. An urban forester might work for a city or **nonprofit organization**. No matter where they work, these professionals share a love of our planet's forests!

Maybe one day, the forest will be your office!
What will you discover?

GLOSSARY

density—the quantity of something per unit of area.

deploy—to spread out, utilize, or arrange for a deliberate purpose.

determine—to make a decision about something using research or observation.

diameter—the distance across the middle of an object, such as a tree.

flammable—easily set on fire.

habitat—a place where a living thing is naturally found.

invasive species—an organism that spreads to an environment where it is not native and causes harm.

monitor—to watch, keep track of, or oversee.

nonprofit organization—a business that benefits the public, or works for the good of a specific cause or group of people, instead of making money for its owners.

prune—to cut away or remove what is extra or unwanted.

sensor—an instrument that can detect, measure, and transmit information to a controlling device.

software—the written programs used to operate a computer.

specialize—to concentrate one's efforts in a specific area.

sustainable—relating to a method of using a resource so that the resource is not used up or damaged.

terrain—the natural features of an area of land, such as mountains and rivers.

vegetation—plant life, especially those covering a certain area.

vulnerable—able to be hurt or attacked.

ONLINE RESOURCES

Booklinks
NONFICTION NETWORK
FREE! ONLINE NONFICTION RESOURCES

To learn more about STEM in the forest, visit **abdobooklinks.com**. These links are routinely monitored and updated to provide the most current information available.

INDEX